DRAGON
COLORING BOOK
FOR ADULTS

THIS BOOK BELONGS TO

_____

COLORIST MANUAL

STEP 1: PUT DOWN YOUR DEVICES, GRAB YOUR COLORING TOOLS, AND BE PRESENT. OPEN YOURSELF TO THIS CREATIVE EXPERIENCE.

STEP 2: SLOW DOWN. BREATHE DEEP. AS YOU COLOR THESE DRAGONS, LET TENSION FADE AS IMAGINATION IGNITES THROUGH THE PROCESS.

STEP 3: ENHANCE DRAGON IMAGES WITH COLOR AND TEXTURE. EXPRESS YOUR CREATIVITY FREELY ON THE PAGE – NO LIMITS, NO JUDGEMENTS. LET YOUR INNER FIRE FLOW.

IMPORTANT: WITH MARKERS, PUT BLANK PAPER BEHIND THE PAGE TO PREVENT BLEED-THROUGH.

THROUGH COLORING'S ESSENCE, CREATIVITY ARISES. STRESS FADES. AN INNER CALM AWAKENS. WITH EVERY PIGMENT-FILLED STROKE, EMBRACE YOUR INNER DRAGON.

AND NOW...EXHALE SLOWLY. YOUR COLORING ADVENTURE BEGINS. DIVE IN AND LET YOUR DRAGON SOUL TAKE FLIGHT!

TRY YOUR COLORS

BEFORE COLORING, USE THIS PAGE TO TEST YOUR
MATERIALS. WITH MARKERS, PUT BLANK PAPER BEHIND
THE PAGE TO PREVENT BLEED-THROUGH.

SHARE YOUR EXPERIENCE

THANK YOU FOR BREATHING NEW LIFE INTO THESE DRAGON IMAGES WITH YOUR CREATIVE COLOR CHOICES AND IMMERSIVE COLORING ENJOYMENT.

WE'RE GRATEFUL YOU TOOK TIME TO SLOW DOWN AND MINDFULLY COLOR THESE MYSTICAL BEASTS. MAY THE CALM FOCUS THIS PRACTICE REQUIRES CONTINUE NURTURING YOUR INNER FIRE AND IMAGINATION.

IF FEELING INSPIRED, SCAN THE QR CODE TO DESCRIBE YOUR COLORING EXPERIENCE. EXPRESS HOW GETTING LOST IN VIBRANT CREATIVITY CONNECTED YOU TO YOUR INNER DRAGON. YOUR STORY COULD SPARK FELLOW ART ADVENTURERS TO UNLEASH THEIR DRAGON SPIRITS THROUGH CREATIVITY AND COLOR.

Made in the USA
Las Vegas, NV
06 November 2024

11208871R00063